ANIMAL COURTSHIPS

edited by
Roger Caras

foreword by
Roger Tory Peterson

A Media General Publication, Richmond, Virginia

Prepared in cooperation with
Photo Researchers, Inc.,
New York, New York.

SBN 0-87858-016-6

INTRODUCTION

The controversy concerning how much about ourselves we can learn from animal behavior, or how much we can learn about animals from our own behavior, will not likely end in this century. But we don't need a resolution to help us appreciate the incredibly complex world of behavior as exhibited by the reproductive cycle.

It is not anthropomorphic to establish a parallel between man and penguin. Even though they both may bring a pebble to their beloved to earn favor, we know that man—neither in body nor behavior—has evolved from the penguin. And so far no one has traced the penguin back to man.

There are, however, striking parallels, appealing contradictions, moments of charm and touching warmth, and some hilarious nonsense, for animals, like human beings, court each other seeking the favor of sex and the assurance of immortality.

Roger Caras
East Hampton, N.Y.

FOREWORD

Most animals have an intricate language other than the sounds they make, a language of movement and display. Colors, patterns and adornment may be used as signals to others of their kind, especially at breeding time, when males assert their dominance over other males, proclaim and hold down property rights, and attract prospective mates. It is tempting to draw human parallels, but the universal rule is that each species has a ritual or set of signals uniquely its own, even though the shades of difference between closely related species may appear very subtle to our eyes. Each signal, whether by voice, movement or adornment, acts as a releaser to an appropriate reaction. The sequence of these acts is often so ritualized that if things do not proceed in a one, two, three manner; if any one of the steps is omitted, mating may not be consummated. This formality, evolved over eons of time, ensures that the species will remain reproductively isolated and stable. An elephant will always mate with an elephant, a rhino with a rhino. There will not be hybrids sporting both tusks and horns roaming the veldt. What would we call them—Rhinophants perhaps?

The displays of some birds are beautiful to watch and in none do they reach such bizarre development as in certain "arena birds" such as the ruff, the cock-of-the-rock, birds of paradise, turkeys, some grouse and pheasants—birds of fantastic adornment. The males dance or display in competition with each other. The females make the ultimate choice, several often choosing the same lucky male.

Most courtship rituals are less spectacular than those of arena birds but very crucial to the survival of the species whether it be giraffe, swallowtail butterfly, honeybee, robin, or rainbow trout.

Roger Tory Peterson
Old Lyme, Connecticut

Rock Pigeon

We have a strangely personal reaction to animals during courtship. After all, an animal displaying for his mate, or actually mating with her, is performing a perfectly normal biological function —like eating or sleeping. Yet, we certainly don't react to animals eating or sleeping the way we react to their courting. Do we react so strongly because we are projecting our own feelings onto them, or because we are embarassed? For whatever reason one can suggest, we are fascinated by animals during their courtship. I have known noisy people become unnaturally quiet—even a little respectful, perhaps—just because a bird spread his tail and began a little dance.

Chameleon

One of the most important functions of courtship is that it keeps similar but different animals from breeding with each other. No two known species of animals have exactly the same courtship pattern. And the minute one animal fails to give exactly the right response a chain of interacting forces snaps, and the animals are sexually incompatible. Imagine what the world would look like if animals could interbreed indiscriminately! A bug-eyed monster from outer space would be a welcome relief.

Lion

"When several males are after the same lioness, arguments are, of course, very common, and if the suitors know each other well enough to be able to judge their opponent's strength . . . the fight may be harmless enough, resulting in no more than a few bloody scratches, because the weaker of the two knows when to give in and retreat. I think that fights for life and death . . .

usually take place between two lions who are strangers to each other. In the Mara area a young male, about four years old, who was too sure of himself, was so badly mauled by a stronger and larger opponent, that he died a few days later."

From *Simba* by C. A. W. Guggisberg
(Bailey Bros. & Swinfen, London, 1961)

When the challenges between males, the displays for the females, the claiming of territory are all over with, there remains, for many species, tenderness. It is very difficult not to attribute human qualities to animals when we see a mated pair being tender, and perhaps even a little coy, with each other. But, we really shouldn't think of them as imitating us. Perhaps, on those rare occasions when we are really tender, we are acting like albatrosses instead of the other way around.

Waved Albatross

Ruppell's Griffon

We must not be subjective when we think in terms of "attractive" and "unattractive." To a gentleman vulture a lady vulture is just as lovely as she can be—and even lovelier if she is coy and offers some resistance. Coyness, even in such "unattractive" creatures as the carrion eaters, apparently stimulates the male. He is encouraged and led on by it, and there may be selective value to it that we do not exactly comprehend. However, our comprehension is unimportant. What is important is that vultures find pleasure in each other's company with male intensity and female coyness being a part of the essential interaction.

Egret (Louisiana)

"Just as the seasons swing back and forth between the bleak sterility of winter and the green fruitfulness of summer, so the lives of birds oscillate between the relative quiescence of the resting season and the impetuous excitement of the breeding season. The extravagant courtship displays that seem so absurdly exaggerated to human eyes are in part a reflection of the intense living of birds and in part a reflection of special needs resulting from their ways of life. . . .

Egret (Spain)

"Mate selection, or pair formation, is one of the functions of courtship in most species, but by no means the only one. At times the songs and displays of courtship serve also to warn away intruders and competitors from the owner's territory. Another important function of courtship is the stimulation of ovulation."

From *The Life of Birds* by Joel Carl Welty (Alfred A. Knopf, N.Y., 1963)

Frigate

Very often animals have special devices or structures that aid in courtship displays. These can be luridly colored bare patches, combs, pouches, parts of the body that can be pumped up with extra blood or inflated with air. Even if the devices are present all year round, they are generally much larger or more colorful during the courtship period. It is not that the potential mate of the displaying bird finds the special equipment *beautiful*—that, of course, is a subjective human reaction—they find it stimulating. Isn't the concept of and the appreciation of beauty a function of stimulation?

Marabou Stork

Jumping Spider

"Male jumping spiders perform extremely droll dances. . . . when the male jumping spider spies a female, it abandons its usual creeping movement. Instead, it raises itself high, beckons with its first pair of legs, and in this erect position runs in zigzags back and forth in front of the female, until at last it pauses and stands face to face with her. If the female remains peaceful and has not stirred, the male can venture to leap upon her and insert his semen-filled palp into her genital orifice."

From *The Sex Life of the Animals* by Herbert Wendt (Simon & Schuster, N.Y., 1965)

Elephant Seal

". . . every male is torn between two conflicting purposes—mating with any and all females he is able to, and driving off or defending himself against other males. It is difficult to do two different things at once successfully.

"A big bull keeps other males away by sheer weight and fighting power. He blows his horn with a roaring snort, the snout inflating and vibrating against the roof of his mouth as the air is driven up. The noise comes from the nose and lips, not the voice at all—far better than anything the Bronx has ever made. If the offending male fails to run at once, the bull charges, often leaving a wake of somewhat injured females and squashed youngsters behind him."

From *Sex and the Nature of Things*
by N. J. Berrill
(Dodd, Mead, N. Y., 1953)

Lion

Generally, love-making among lions is less chaotic than among other cats. There are battles between competitive males, of course, and there is always the possibility that a male and a female will have a misunderstanding, which can be titanic when it does occur, but usually the affair is pretty quiet. The male follows the female everywhere she goes during her four-day heat and rolls his lips back and makes faces. And they mate very often—a hundred or more times in the four days, in some cases. The male lion is very tired by the time the female's heat has passed.

Blue-faced Booby

"For bisexual animals, a necessity for survival of the species is the discovery of a suitable mate. Recognition of a receptive partner is the first link in the chain of events leading to union of the sex-cells. Far from being a random search, mate-finding in higher animals is a highly organised process which involves one or more of the senses of sight, smell, sound, touch and even taste."

From *Animal Behaviour*
by J. L. Cloudsley-Thompson
(Macmillan, N.Y., 1960)

Peacock

In our attempts to attribute human characteristics to animals we have assigned vanity to the peacock. When the male peacock spreads his astounding tail and struts we say he is vain; when we strut we are peacocks. The comparison is a false one. Not only can we not be peacocks, but when the peacock struts he is not exhibiting vanity. He is pleading, bursting with a drive, a need that is as old as life itself. If he were actually trying to prove his uniqueness, *that* would be vanity. But he is simply playing the mating game. What he is saying, really, is, "Please, pick me. I am the ideal peacock and you will have ideal chicks by me."

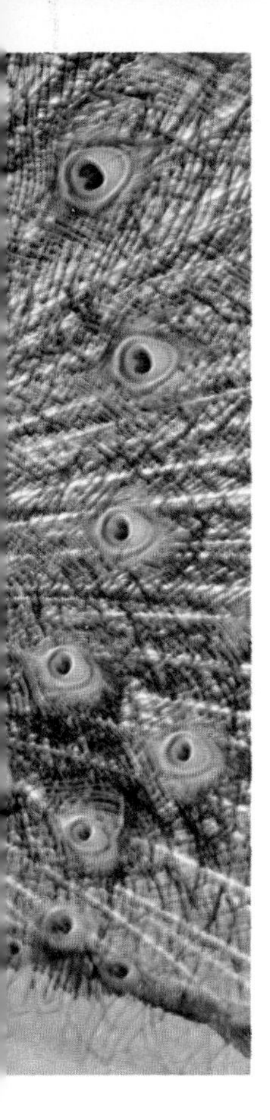

Humbolt's Penguin

"Penguin displays are mostly variations on a basic posture involving stiffly outstretched flippers and vertically pointing bill. . . . One of the displays prominent in a rookery in the breeding season is called the 'Ecstatic.' . . . The flippers are moved in a rhythmic way, while the chest is thrust outwards, the bird giving vent to a drumbeat-like cry. It reaches the climax as the head feathers are raised and the call changes to a braying sound."

From *Penguins*
by John Sparks and
Tony Soper
(Taplinger, N.Y., 1967)

Common Egret

The incredibly beautiful plumage some birds develop for their courtship season has led man to some terrible excesses. Our lust for egret feathers nearly resulted in that bird's extinction and did end up with men actually killing other men. Ironic, isn't it? When ladies don the feathers of the egret they are reversing the natural process. In the natural world the male egret grows those feathers to please and attract the female. Therefore, to set things right, men will have to wear the feather boas in the future.

Short Horned Owl

Mating among owls can be hazardous. Except for the snowy owl, and then only to a limited degree, male and female owls tend to be identical—although females are generally larger. Sex recognition, then, becomes a dangerous business. A bad guess could put a sexually aroused male owl into the clutches of a rival interested only in an expanded hunting range. The lack of sexual dimorphism is compensated for by behavior. Male and female owls act and sound differently when it is time to court. The difference between a threat and a "come-up-and-see-me-sometime" look may be vague to us but not to owls.

Manatee

It is easy to watch graceful and splendidly adorned animals brushed by the mood of love and comprehend what it is all about. However, when the animal is a twelve-hundred-pound amorphous sack of blubber with a pendular lip and beady eyes, when he is covered with algae and submerged in dirty water, it is difficult to conjure up any image of love. To a manatee, however, our observations are of little importance. That's why there are little manatees.

Fiddler Crab

"The little fiddler crabs, *Uca,* are characterized in the males by the possession of one relatively enormous claw which is normally carried close to the body. It is a formidable-looking weapon, but seems never to be used except in combat with other males of the same species and as a secondary aid in digging. . . . At breeding time, and particularly when a female is nearby, the males make a peculiar gesture, extending the big claw to its full length and then whipping it suddenly back toward the body. One might conclude, with some logic, that the big claw is a distinctly sexual attribute comparable to the horns of a stag."

From *Between Pacific Tides*
by Edward F. Ricketts and Jack Calvin
(Stanford University Press, Stanford, 1962)

Kissing Gouramis

What are the kissing gouramis above doing? Kissing, apparently. Is this part of their mating ritual? We are not at all sure. We don't know why gouramis kiss, but we do know they go around kissing other fish on the body—fish they could not possibly mate with—and it appears to be very annoying.

Grunion

There is enormous diversity in the way fish court (when they court) and reproduce. The remarkable grunion is washed up onto the beach and there the whole process takes place—in the open air. Each female may have one or more males wrapped around her as she lays her eggs in the sand. The eggs are quickly fertilized and the fish wriggle back into the water. The next incoming wave covers the eggs with sand.

Siamese Fighting Fish

Siamese fighting fish have been domestically bred for a very long time and have been developed into truly gorgeous creations with great flowing fins and neon-like colors. Although there has been an endless amount of nonsense written about these splendid fish, it *is* true that the males will fight violently when exposed to each other. They can only be kept one to an aquarium, and some males are so full of hostility they can't even be used for mating; they kill their mate, or eat their eggs. They can be said to be somewhat overdeveloped in the hostility department. Other fish don't find it necessary to act that way.

Yellow-headed Jawfish

It is interesting to see how some species carry through the promises made during courtship. There are about a dozen species of jawfish now known and they build tubes or tunnels in the sand; the males can be aggressively territorial. Apparently all species carry or "incubate" the eggs in their mouths, which might explain the oversized jaw. It is not clear if it is the male or the female that so devotes itself to the next generation. This may vary from species to species.

Siamese Glassfish

"The annual inundation of the vast central plains of Thailand and of the various minor plains is an event of great importance in the life of all the fishes. As the streams begin to rise and fill their beds, together with the connecting canals, and the tributary ponds, lakes, swamps and marshes that had become reduced during the protracted dry season, the fishes follow the flood-waters, into the rice fields, into the lakes, and into the swamps being converted into lakes, and by the time the inundation has reached its height the vast majority of the free-swimming fishes has spawned."* And so the sex play, the display and the competition, and ultimately the reproduction of the fish is tied to the moon, to the sea, to the wind, and to the rotation of the planet.

*From *The Fresh-Water Fishes of Siam, or Thailand* by H. M. Smith (U. S. National Museum Bulletin, 1945, *18*)

Kissing Grunt

The form of the fish, its color, its sometimes bizarre appearance are the results of evolution. The end result we witness is, at this point in the evolutionary development of the fish, the best way for it to look if it is to survive. This is equally true of its behavior. According to the precepts of ethology—a science of our time—the behavior of the fish, its courtship rites, also derive from a long and arduous evolutionary history. It is really impossible, then, to understand appearance or behavior without the other, for neither evolved in a vacuum. The fish' appearance and its behavior are interlocking elements and represent the interplay of an organism with its environment—and environment, in this sense, refers to its ability to reproduce itself—and that includes social structure. Courtship is very difficult to extract from a very grand whole.

Fringed Filefish

Honey Bees

". . . the workers have been left without a queen, but with the promise of one or more developing in the queen cells. The first young queen to emerge is usually destined to become the new queen of the old colony. She searches out the other queen cells and she, or the workers, or both, destroy all the other young queens. Within a few days the young virgin queen flies out of the hive on her mating flight. . . . As she flies she is pursued by drones. One of these eventually overtakes her and they mate. . . . So firmly are portions of his genital system held in place in the queen's body during and following the mating that she can free herself from him only by ripping them from his body. The male quickly dies. . . ."

From *Lives of Social Insects* by Peggy Pickering Larson and Mervin W. Larson (World, N.Y., 1968)

Praying Mantis

When we watch nature we must admit there are some things that are more than just a little puzzling. Take the mantis—take any of them, there are about eighteen hundred different kinds. Their very name means soothsayer, which suggests wisdom. According to the Moslems, these insects are fastidious when they pray—they always face Mecca. Considering that, you might think they are very well-adjusted insects. But they aren't; they have a quirk. The lady mantis, when she is done with her lover, and sometimes even before she is done, eats him. It is rough on him, and it casts a shadow on her in the eyes of most people.

Damselfly . .

Damselflies (a kind of dragonfly) are a little more considerate of each other than the mantids. The pair mates either while perched or in flight and then, after all the essentials are over, the female takes the male for a ride. He rides her tandem fashion to the water plant where she will deposit her eggs. He hasn't anything to do at this point, but she still carries him around and lets him watch their eggs being put into position. Taking your mate for a ride seems nicer than gobbling him up.

In courtship—in all aspects of sexual reproduction—the sense of smell is often very important. With no group of animals is this more true than with moths. In the case of the emperor moths, for example, the female discharges an infinitesimal quantity of sex attractant from glands near the hind tip of her body. Males gather from astonishing distances to acknowledge her message and participate in immortality. The amount of scent required to attract males from hundreds and perhaps thousands of yards is so minute as to be almost unmeasurable. When you consider the volume of scent and the volume of air into which it is discharged you realize its incredible power. The males have enlarged antennae with vastly increased surface area to facilitate participation.

Emperor Moth

Sage Grouse

". . . the actions of a single cock . . . paying court to several females near him. . . . His large, pale yellow air sacs were fully inflated . . . giving the bird an exceedingly peculiar appearance. He looked decidedly top-heavy and ready to topple over at the slightest provocation. The few long, spiny feathers along the edges of the air sacs stood straight out, and the grayish white of the upper parts showed in . . . contrast with the black . . . breast. His tail was spread out fan-like . . . and was moved from side to side with a slow quivering movement. The wings were trailing on the ground . . . he moved with short, stately, and hesitating steps . . . uttering low, grunting, guttural sounds. . . ."

From *Life Histories of North American Birds* by Charles Bendire (Smithsonian Institution Special Bulletin #1, 1892)

"Pre-coitional behavior consists of dipping the head deep under the water and then lifting it and at the same time throwing water over the back with the back portions of the head and neck, in a manner similar to the movements of bathing. But, unlike ordinary bathing, the wings are not used; they are kept folded in their normal resting position. This . . . display, which may be initiated by either the male or female, gradually increases in frequency and intensity . . . the male works closer to the female, and typically grasps the feathers of the back of her neck. . . ."

From *Social Behavior and Breeding Success in Canada Geese (Branta Canadensis) Confined Under Semi-Natural Conditions* by N. E. Collias and L. Jahn *(Auk,* 1959, 76(4))

Giant Canada Goose

Mallard Duck

"... reproductive behavior in ducks may be conveniently divided into an early phase of conspicuous displays associated with actual pair formation, followed by the later and less elaborate behavior patterns concerned with pair bond maintenance and fertilization. Two possible advantages of the considerable time lag between pair formation and egg laying are that it decreases the likelihood of uncorrected mismatings between species and, furthermore, provides the female with the protection of a mate to ward off unmated males that might attempt to rape her."

From *The Evolution of Duck Courtship*
by Paul A. Johnsgard
(*Natural History Magazine*, Feb., 1968, 77(2))

Iguana

The true iguanas are often large and formidable lizards. They look more like dragons, and if they were big enough that is probably what we would call them. Still, coyness plays a big role in their courtship. When she is certain a prospective mate is sufficiently lathered up, really inflamed, the female iguana leads him on a merry chase. While making small physical signs of her readiness to mate she dashes off tail high and taunts him until his passion can no longer be contained. This behavior would certainly tend to eliminate the half-hearted suitor and probably goes a long way toward assuring mating success.

The sounds animals produce often play an important part in courtship. This is obvious among birds, but it is no less so among the frogs and toads. The spring peeper is an inch or less in length, yet his mating calls fill the night air and can be deafening at the edge of a swamp. When the little male inflates his throat sac it gleams like a white bubble—it is actually transparent—and can be half the size of the animal itself. The volume produced is astounding and very pleasing to lady frogs nearby. It carries a message they can hardly fail to attract.

Spring Peeper

Despite our long-running interest in the way animals reproduce themselves, we are occasionally outwitted. Here is an interesting case. The chuckawalla is the second bulkiest lizard in North America—only the Gila monster has more body volume—yet, aside from the fact that we know it lays eggs, we know nothing about its courtship and breeding behavior in general. It has so far managed to elude us during the more intimate phases of its life.

Chuckawalla Lizard

Doves

"Every full-grown male pigeon is on the alert to find a mate. For days he goes courting, cooing, and dancing before many birds in the hope of chancing upon the unattached female that will join him in housekeeping. Since he cannot tell the sex of other pigeons on sight, he must through trial and error find out which birds are likely to resist his advances with cooing and fighting—males like himself—and which birds will only peck at him indifferently and edge away as if afraid . . . our courting male is bound to make many a serious blunder before he chances upon a female that will tolerate his eager presence. . . ."

From *The Book of the Pigeon and of Wild Foreign Doves* by Carl A. Naether (David McKay, N.Y., 1964)

Ruffed Grouse

"Long before dawn the birds takes his station and struts back and forth, spreading his immaculate fantail in a display pose. Then, thrusting himself upright . . . the cock bird sounds his challenge to other males and to lovesick hens within hearing. Nothing in nature is quite so elfin and mysterious as the drumming of a grouse. It is curiously ventriloquial and difficult to locate—a series of resounding thumps that accelerate into a sudden blur of sounds. . . . A hen always demurely approaches a drummer. Often she is rebuffed at first contact. Later she repays his arrogance by resisting his advances and compelling him to follow her like any lovesick suitor."

From *The Partridge Stands Pat* by Frank Woolner
(*True,* November, 1971)

Hoopoe

The hoopoe, which has the delightful scientific name, *Upupa epops,* apparently mates for life, or at least it spends all of its time in company with a member of the opposite sex. When it comes time to breed the male begins short little intimate ceremonies. He presses his slender beak against a branch, erects his splendid crest, and bows repeatedly to his mate making a funny little cré cré cré noise.

I know of nothing in the long catalog of animal behavior more astounding than the construction of the bower by the bowerbird. Built by the male, the bower is elaborate and intricately decorated and completely independent of the nest which may be built later in the season and hundreds of feet away. Several species of bowerbird even paint their structures. This is nothing less than amazing. We assume the construction of the bower is part of the courtship and mating ritual.

Satin Bowerbird

Capercaillie

"A cock capercaillie in full display is a wonderful sight. He droops his wings, expands his tail, which he raises almost vertically above his back, and stretches up his neck. During part of his 'song display' his throat feathers stand out rather like those of an angry raven. . . . Two cocks sometimes display at one another—in the posture described—and jump into the air in front of a single hen which appears to ignore them . . . cocks do not fight seriously in these ritual combats. . . ."

From Notes by Desmond Nethersole-Thompson
quoted in *The Birds of the British Isles*
by David Armitage Bannerman, Vol. 12
(Oliver & Boyd, London, 1963)

Black-footed Albatross

Courtship is as much a part of evolution and as critical to the future of a species as the physical devices that have undergone evolutionary development. Usually courtship is the province of the male. He is often ready to breed before the female, and it is believed that his displaying can have a great deal to do with her catching up physiologically. In some species, however, display is mutual. Male and female together engage in the songs and the dances, the feeding and the preening that together lock their species into a continuum and assure immortality. In a way we could say that is what courtship is all about—immortality.

PHOTO CREDITS

PAGE	SUBJECT	PHOTOGRAPHER
5	Rock pigeon	Toni Angermayer
6	Chameleon	William M. Stephens
8	Lion	Ian Cleghorn
9	Lion	Ian Cleghorn
11	Waved Albatross	George Holton
13	Ruppell's Friffon	R. D. Estes
14	Egret (Louisiana)	Lewis W. Walker
15	Egret (Spain)	Roger Tory Peterson
17	Frigate	George Holton
17	Marabou Stork	Russ Kinne
18	Jumping Spider	J. H. Carmichael, Jr.
19	Jumping Spider	J. H. Carmichael, Jr.
20	Elephant Seal	Tom McHugh
21	Lion	Louise Bucknell
23	Blue-faced Booby	James Murray
24	Peacock	Moos-Hake/Greenberg
25	Humbolt's Penguin	Private Collection
26	Common Egret	Russ Kinne
27	Short Horned Owl	Roger Tory Peterson
28	Manatee	Russ Kinne
29	Fiddler Crab	Jane Burton
30	Grunion	Tom McHugh
31	Kissing Gouramis	Russ Kinne
32	Siamese Fighting Fish	Jane Burton
34	Yellow-headed Jawfish	Russ Kinne
35	Siamese Glassfish	Jane Burton
36	Fringed Filefish	Jane Burton
37	Kissing Grunt	Carleton Ray
39	Honey Bees	Colin G. Butler
40	Praying Mantis	Jen and Des Bartlett
41	Damselfly	Russ Kinne
42	Emperor Moth	Jen and Des Bartlett
44	Sage Grouse	Bill Gabriel
44	Sage Grouse	Joe Van Wormer
45	Sage Grouse	Joe Van Wormer
45	Sage Grouse	Bill Gabriel
46	Giant Canada Goose	Karl H. Maslowski
48	Mallard Duck	Mary M. Thacher
48	Mallard Duck	Joe Van Wormer
51	Iguana	Toni Angermayer
52	Spring Peeper	Russ Kinne
53	Chuckawalla Lizard	E. Stanford
55	Doves	Jerry Cooke
56	Ruffed Grouse	Bill Gabriel
58	Hoopoe	Toni Angermayer
59	Satin Bowerbird	Toni Angermayer
60	Capercaillie	Ake Wallentin Engman
62	Black-footed Albatross	George Laycock